MAKE it WORK!
SOUND

Andrew Haslam

written by
Alexandra Parsons

photography by
Jon Barnes

World Book

in association with
WCN

MAKE it WORK!
Other titles

Body
Building
Dinosaurs
Earth
Electricity
Flight
Insects
Machines
Photography
Plants
Ships
Space
Time

Published in the United States by
World Book Inc., 233 N. Michigan Ave., Chicago, IL 60601
in association with Two-Can Publishing.

2001 printing
Copyright © 2001, 1997, 1992 Two-Can Publishing
Design copyright © 1992 Wendy Baker and Andrew Haslam.

**For information on other World Book products,
call 1-800-WORLDBK (967-5325), or
visit our Web site at http://www.worldbook.com**

Library of Congress Cataloging-in-Production Data
Haslam, Andrew
    Sound / Andrew Haslam; written by Alexandra Parsons; photography by Jon Barnes.
        p. cm. — (Make it work!)
    Previously published: New York: Thomson Learning, 1995.
    Summary: Introduces basic facts about sound, how it is produced, transmitted,
and received with instructions for related experiments and projects.
    ISBN 0-7166-4705-2. — ISBN 0-7166-4734-6 (pbk.)
    1. Sound—Juvenile literature. 2. Sound—Experiments—Juvenile literature.
[1. Sound—Experiments. 2. Experiments.] I. Parsons, Alexandra. II. Barnes,
Jon, ill. III. Title. IV. Series.
    QC225.5.H37   1998
    534'.078—DC21   98-13425

Printed in China

(hc) 4 5 6 7 8 9 10 09 08 07 06 05 04 03 02 01

Editor: Mike Hirst
Illustrator: Michael Ogden
Additional design: Belinda Webster

Thanks also to: Albert Baker, Catherine Bee,
Tony Ellis and everyone at Plough Studios.

Contents

Words marked in **bold** in the text are
explained in the glossary.

Did you know that sounds can blow out candles and crack glass? Or that bats use sounds to "see" and find their way around? Have you ever wondered how different musical instruments produce completely different sounds?

MAKE it WORK!
Start experimenting with sound. Find out how sound waves are made, how they travel, and some of the surprising things they do.

You will need
Most of the activities in this book use simple equipment, such as cardboard, glue and odds and ends. However, you will find some specialist equipment useful.

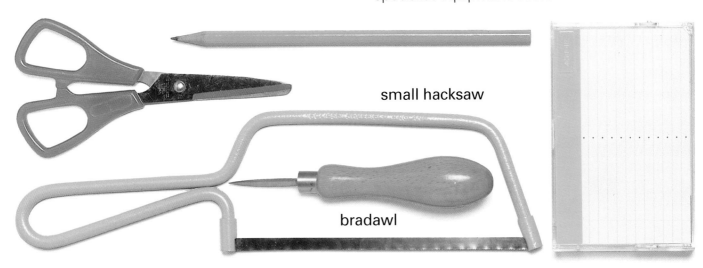

small hacksaw

bradawl

Investigating sound is part of the science of **physics**. Sound is a kind of energy that flows through the air in invisible waves. Everything that makes the air move, from the rustling of a leaf to the pounding of a big bass drum, sets off **sound waves**, and creates a different kind of sound.

Small hacksaw and bradawl Some of the activities include making simple wooden supports or structures. You should always ask an adult to help if you are using saws, utility knives or other sharp tools.

Cardboard tubes These are handy for making musical instruments. You can use the insides of aluminum foil rolls or cardboard tubes made for carrying posters.

Music manuscript paper This is useful for writing down musical notes.

Copper wire This will be needed for making a simple radio and a telephone. Sound waves can be turned into an electrical pulse and then transferred along the wire. You can buy copper wire at any hobby shop or electrical supplier.

Balloons and rubber bands Some of the projects involve making a drum. Rubber from a balloon makes an excellent drumhead, and elastic bands will hold the rubber in place without tearing it.

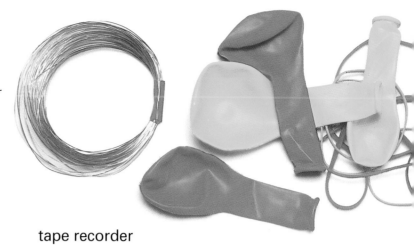

tape recorder

earphones

Tape recorder A simple tape recorder is the most useful piece of equipment for sound experiments. Ideally, you will need a small tape recorder or personal stereo, along with a microphone, small headphones and some blank cassettes.

Plastic tubing and corks These can be bought from hobby shops.

microphone

When we hear a noise, the sound waves usually travel to our ears through air, but sound can move through other substances too. Whales, for instance, hear a wide range of sound waves that move through water.

6　Sound Waves

We hear sound when a moving object makes the air **vibrate**. These vibrations travel through the air in the form of waves, and are picked up by our ears as sounds. The shape of the sound wave depends on the **pitch** of the sound. Low-pitched sounds are deep and rumbling, like a big bass drum. High-pitched sounds are shrill and piercing, like a whistle.

MAKE it WORK!

If you could see air, it would look like a huge floating soup of gas particles. Low noises would make ripples that are far apart, and high-pitched sounds would make waves that are very close together. In fact, sound waves in air are invisible, but you can certainly prove they exist. Here are two ways to observe their effects.

For the sugar drum you will need

a cake tin	sugar
a wooden spoon	a balloon
large rubber bands	a baking pan

1 Cut out a circle of balloon rubber. Stretch it over the cake tin and secure it with rubber bands.

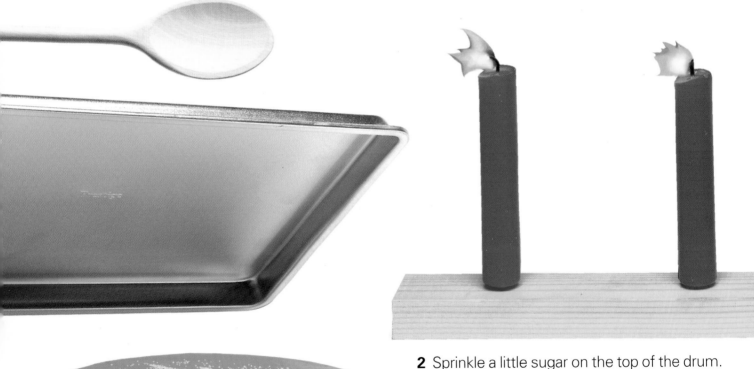

2 Sprinkle a little sugar on the top of the drum.

3 Hold the baking pan above the drum and hit it with the wooden spoon. As the sound waves reach your ear, you hear the sound of the spoon on the pan. When those same waves hit the drumhead, they make it vibrate and you can see the sugar dancing up and down.

◀ Try holding the baking pan closer to the drum and then farther away from it. Does this affect how much the sugar moves?

For the sound cannon you will need

a cardboard tube   candles
a piece of plywood   long nails
a plastic bag or a balloon   rubber bands

1 Ask an adult to help you hammer three nails through a piece of wood. Turn the wood over and push a candle onto each nail.

2 Take a piece of cardboard tube. The inside of a roll of aluminum foil will work well too.

3 Stretch a circle of balloon rubber or plastic bag over each end of the tube and secure each with rubber bands.

4 Make a little hole in the plastic stretched over one end of the tube.

5 Light the candles.

6 Point the end of the sound cannon with the hole in it at one of the candles. Hold it just a few inches away.

7 Tap the other end with your finger. The vibrations you make by tapping the drumhead travel to your ear as sound waves. The same vibrations move down the tube and push the air through the little hole at the opposite end, blowing out the candles.

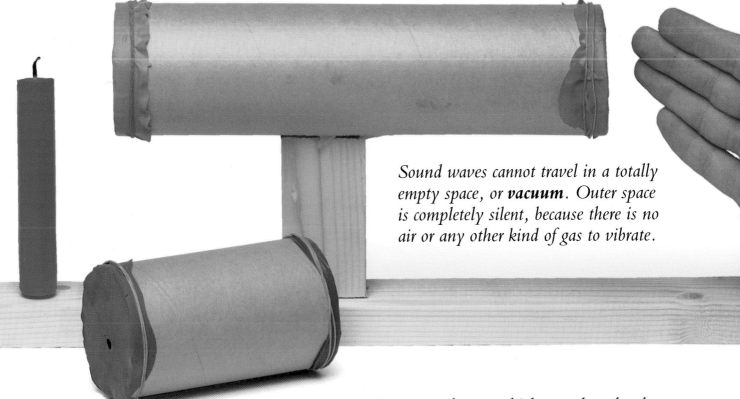

*Sound waves cannot travel in a totally empty space, or **vacuum**. Outer space is completely silent, because there is no air or any other kind of gas to vibrate.*

*Scientists measure sound waves by the number of times they make the air vibrate each second. One vibration, or cycle, per second is called one **Hertz** (Hz). When you hear a 700 Hz noise, the sound waves are hitting your eardrum 700 times per second. Scientists also measure other types of waves, such as light waves or radio waves, in Hertz.*

Some sounds are so high or so low that humans cannot hear them. We are unable to make out sounds that have a frequency above roughly 20,000 Hz, or below 20 Hz. However, many animals have a much wider hearing range than humans. Bats and dogs, for instance, both pick up much higher sounds than we do. There are even special dog whistles that give out a high-pitched noise which only dogs can hear.

8 The Ear

Our ears are specially designed to pick up vibrations in the air and change them into **nerve pulses** which our brains then understand as sounds. The working parts of our ears are actually inside our skulls. The flaps that stick out on either side of our heads are just funnels used to collect sounds and pass them along to the **eardrum**.

MAKE it WORK!

Make your own working model of a human ear. The eardrum vibrates, moving three small bones inside the inner ear. These bones in turn move a fluid through a curly pipe called the **cochlea**. The cochlea is lined with minute hairs, and as the fluid moves, so do the hairs, sending tiny sound impulses along the nerves to the brain.

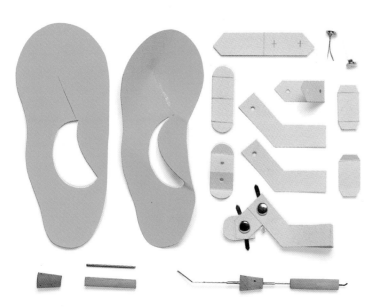

You will need

wood and nails	two cork stoppers
thin dowels	one piece of longer,
rough twine	thinner cork
plastic tubing	thin poster board
paper fasteners	stiff wire, a balloon
a cardboard tube	tape and rubber cement
thin copper tubing	food coloring

1 Take a piece of cardboard tube and cut out part of one side as shown. Place a circle of balloon rubber over one end, fixing it securely with rubber bands. This will be the eardrum.

2 Make the bones that are connected to the eardrum – the **malleus**, the **incus** and the **stapes**. In the model, these bones are made from pieces of poster board. Cut out and fold the shapes shown in the photograph. Then glue and fix them together with paper fasteners as shown in the diagram on the right.

3 Stick the flaps of the cardboard malleus onto the rubber of the eardrum with rubber cement.

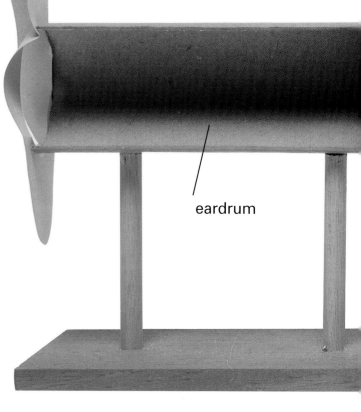

eardrum

4 Cut a piece of flexible, transparent plastic tubing and thread a slightly shorter piece of string through it. Use a rough kind of twine, made out of natural fibers. The plastic tube is the cochlea, and the fibers on the twine are the tiny hairs that send nerve pulses to the brain.

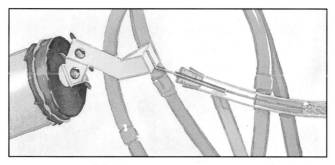

Making the inner ear

5 Connect the "bones" to the "cochlea." Ask an adult to drill a narrow hole in one cork stopper and push a small piece of thin copper tubing through the hole.

Then thread the stiff wire through the copper tubing and stick it into the thinner piece of cork as shown.

6 Push the cork and wire mechanism into one end of the plastic tube. Almost fill the tube with water, add several drops of food coloring, and use the remaining cork to stop up the other end.

7 Make a wooden base and put the separate parts of the model together, holding them up on dowel supports. Glue the wire to the cardboard "stapes" and twist and tape the "cochlea" into a spiral shape.

8 Make an outer ear shape from colored poster board, and glue it to the free end of the tube.

malleus

organs of balance

stapes

incus

cochlea

▲ Ears are important for things other than hearing. They also help us to keep our balance. You can make your model ear even more authentic by adding semicircular tubes filled with colored water. These represent the fluid canals that are our organs of balance.

Operating the model

To watch your model work, tap lightly on the inside of the eardrum to make it vibrate as though it had been hit by a sound wave. Can you see what happens to the little hairs on the string inside the cochlea tube?

Have you ever wondered why you have two ears, placed on either side of your head? It's to help you tell which direction a sound is coming from. Unless a sound comes from directly in front of you, one ear will always pick up more of it than the other. Your brain uses the different information coming from each ear to work out the direction of the sound.

1 Take a large sheet of paper and fold it in half. Attach a pencil to one end of a piece of string and fasten the other end of the string to the piece of paper at a corner along the fold.

2 Use the string and pencil like a compass to draw an arc. Then cut along the line, and unfold the paper to give you a semicircle. Repeat with another piece of paper the same size.

3 Put the two semicircles together to make a whole circle. If you wish, you can divide the semicircles into segments and make alternate colors. Then cut out a paper arrow.

MAKE it WORK!

To find out where a sound is coming from, we usually turn our heads until both ears hear the sound equally and the sound is "in focus." Make this game and test whether your ears have a good sense of direction.

You will need

large sheets of
   colored paper or thin poster board
a pencil, scissors and string
some friends
a blindfold

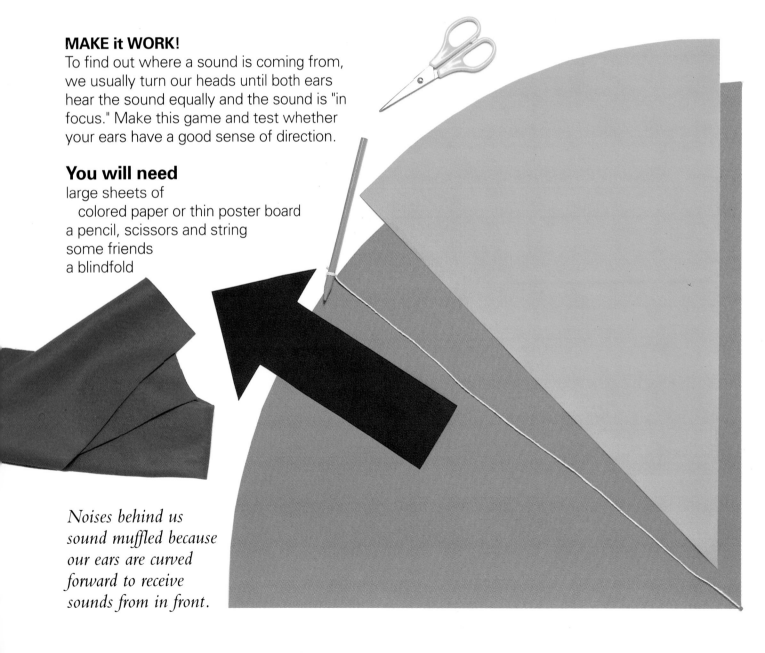

Noises behind us sound muffled because our ears are curved forward to receive sounds from in front.

Compared with light, sound waves travel quite slowly – light moves almost a million times faster than sound. That's why, during a thunder storm, we see a flash of lightning before we hear a clap of thunder, even though they are both, in fact, caused at exactly the same time.

▼ Playing the game

This is a game for three to five people. One person volunteers to wear a blindfold and sits in the middle of the circle. The other players stand or sit silently around the circle, and someone makes a gentle noise, such as clicking his or her fingers. The blindfolded person points the arrow in the direction they think the sound is coming from. Each player can have a turn at being blindfolded. Whose ears have the best sense of direction?

▶ Ask a friend to stand at the other side of a field or playground, holding up a handkerchief. (The farther away your friend, the better the experiment will work – a pair of binoculars could help here.) Tell the friend to shout and drop the handkerchief at exactly the same time. You should see the handkerchief begin to fall before you hear the shout.

Instead of poster board, you could make the circle out of felt or another material.

Children have a wider range of hearing than adults. They can hear higher-pitched sounds.

12 Reflecting Sound

Sounds travel quite fast and far through air, but when they come up against a solid object, their path is blocked – which is why you can hear sounds only faintly through walls. If the solid object is soft, like a cushion, the sound is simply soaked up. But if the object is hard, like a wall, most of the sound bounces off it like a ball off the floor. In fact, if you shout in a large hall with hard, smooth walls, you may actually hear your voice bouncing back off the walls in an echo.

Angles of bounce

Sound waves in air will bounce off a flat, solid object at the same angle as they hit it – just like a bouncing ball off a wall. If, however, the sound waves are bounced off a surface that is soft or bumpy, the waves will break up and fade away.

You will need

thick poster board
a utility knife and ruler
cardboard tubes
a tape recorder and microphone
a clock or watch with a very soft tick
modeling clay
an egg carton

MAKE it WORK!

In this experiment, you can control the path of the sound waves by directing them along cardboard tubes. The tubes hold the sound together, making it louder because the sound waves can't spread out and get lost in the air around them.

1 Take four equal pieces of cardboard tube, and cut three squares out of a piece of thick, smooth poster board.

2 Use modeling clay to secure the cardboard tubes and the squares of poster board in position as shown below. Each tube must be placed at exactly the same angle to the squares as the others have been.

3 Measure the distance in a straight line between one end of the zig-zag and the other.

4 Set up your clock or watch away from the tubes, and record it ticking across the distance you have measured. Your microphone will pick up only a faint sound, or no sound at all.

5 Now position the clock at one end of the zig-zag, and record the sound that comes out of the other end.

6 If all of your tubes are positioned at the same angle, you should be able to record the ticking sound clearly. The sound waves travel down one tube, bounce on and off the reflector card at the end, and continue back down the next tube.

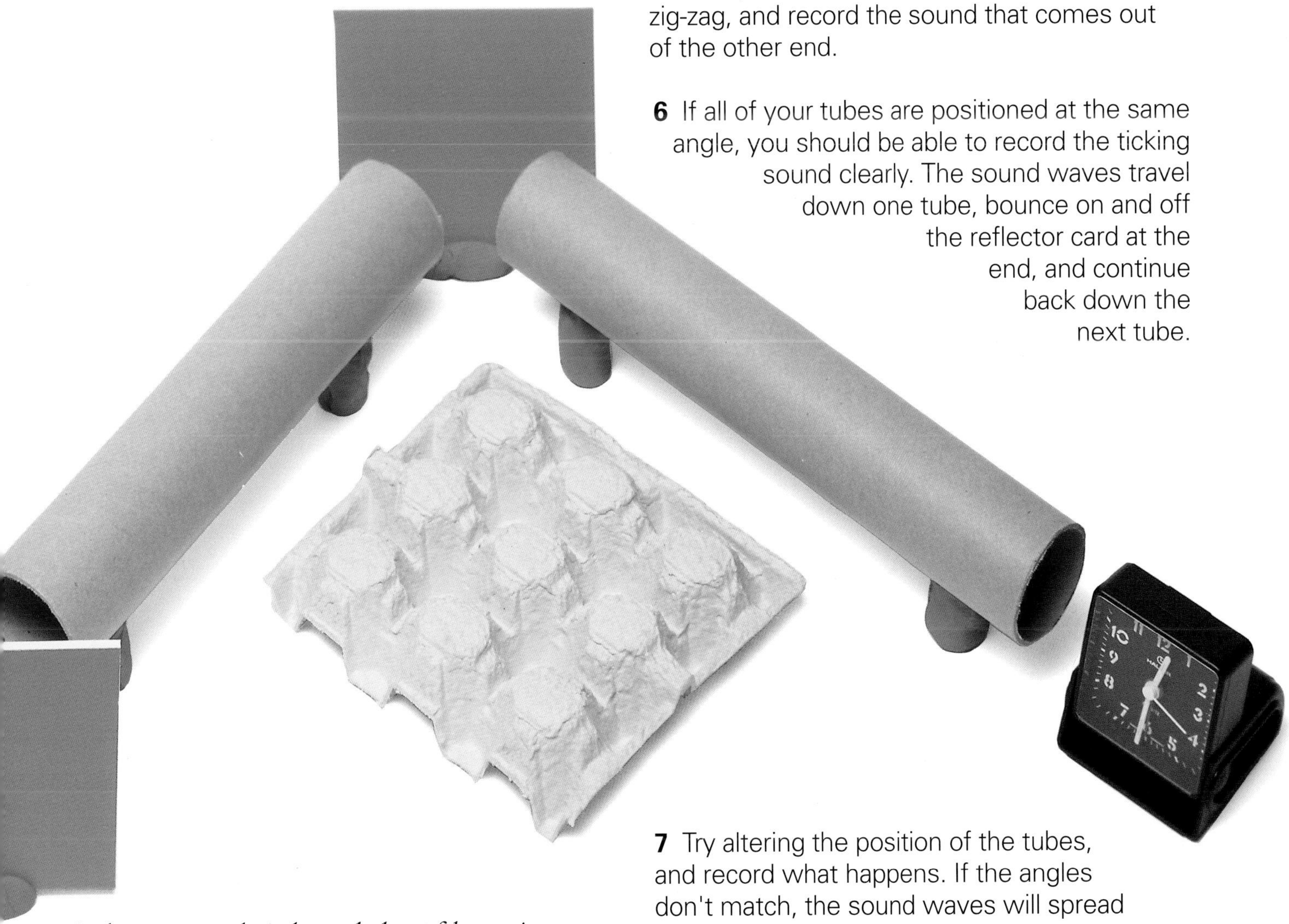

7 Try altering the position of the tubes, and record what happens. If the angles don't match, the sound waves will spread out into the surrounding air, getting weaker and weaker.

Deadening sound
Put the cardboard tubes back in their original zig-zag positions and then experiment with reflector cards made out of different materials. Cut some squares out of an old egg carton, so you can test the effect that a bumpy, uneven surface has on sound waves.

Architects use their knowledge of bouncing sound waves when they design new buildings. A noisy restaurant can be made much quieter by covering the floor, walls and ceiling with soft fabrics and bumpy surfaces to deaden the sound. But the stage and walls of a concert hall can be built to reflect sound waves, so that the music travels clearly toward where the audience is sitting.

14 Amplifying Sound

Before electronic hearing aids were invented, people who had difficulty hearing used an ear trumpet. They would put the narrow end of the ear trumpet to their ear, and if someone spoke clearly into the wide end, they could hear that person's voice more clearly. The ear trumpet **amplified** the sounds, or made them louder.

The simplest amplifier is a big cone. It can be used to send out sounds or to listen to them. When it is used to send sounds, the cone holds the sound waves together, so they don't spread out in the air so quickly. When a cone is used to listen, like the old-fashioned ear trumpet, it collects sound waves from the air and directs them into the ear so they sound louder.

You will need

thin colored poster board   glue or tape
flexible plastic tubing   scissors
two plastic funnels

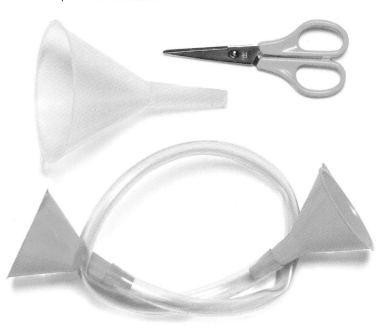

MAKE it WORK!

You can make some instruments to amplify sound waves. Doctors use a special listening device called a **stethoscope** to hear inside their patients' bodies. Normally, we can't hear the quiet gurgles and thumps made by our bodies, but the stethoscope picks up the sound waves and leads them directly into the doctor's ear.

Making a stethoscope

Take a piece of plastic tubing that fits neatly over the narrow ends of the two funnels. Attach a funnel to each end.

Ask a friend to hold one funnel over his or her chest, put the other funnel to your ear and listen carefully. You will be able to hear your friend's heartbeat.

*The loudness of a sound is measured by the force with which the sound wave pushes the air. The units of measurement are called **decibels** – named after the scientist and inventor Alexander Graham Bell.*

Making an ear trumpet and a megaphone

1 Take a sheet of thin, colored poster board and roll it into a cone shape. One end should be wide, to collect sound waves, and the other end should be narrow enough to fit into your ear.

2 Make the megaphone the same way, but the narrow end should be slightly larger so that you can speak into it.

3 Once you have the right cone shapes, tape or glue the poster board. Trim the ends and decorate the cones with colored paper shapes of bright, contrasting colors.

4 To test the megaphone, ask a friend to stand far enough away so you can't hear each other's voices. Then speak normally through the cone.

5 To test the ear trumpet, point it toward a quiet noise. You will be able to hear the sound more clearly. **Be careful!** Never shout at anyone down an ear trumpet. You could damage their eardrum.

On the decibel scale, 0 is absolute silence. A falling leaf would measure 20 decibels, a conversation about 50 and a clap of thunder 110. Above 140 decibels, sounds become painful and may damage the eardrum.

16 Traveling Sound

Our ears usually pick up the sound waves traveling through the air, but, in fact, sound waves can travel through all kinds of substances. Only in a complete vacuum is there silence because there is nothing to transmit the sound. Sound actually travels much faster and farther through water than it does through air. It can even travel fast through a solid object, such as wood, if the sound makes the wood itself vibrate.

MAKE it WORK!

Telephones convert your voice into an electrical signal to transmit it around the world, but even a length of twine will transmit sounds over a short distance more effectively than the air. This is a simple telephone made from just twine and old soup or coffee cans.

You will need

a length of twine, about 19½ ft (6 m) long
two identical soup or coffee cans
a drill (to be used by an adult)
colored paper and tape

household twine

covered cans

1 Ask an adult to drill a hole in the bottom of each can, just big enough for the twine to be threaded through. Cover the cans with colored paper for decoration.

2 Tie a big, thick knot in one end of the twine and thread it through the hole in the can from the top end.

3 Thread the loose end of the twine through the bottom of the other can, then tie a big, thick knot to stop it from slipping back through.

4 You take one can, and get a friend to take the other. Walk apart until the twine is taut.

5 Put the cans up to your ears and move until the twine is as taut as possible between you.

6 Now, keeping the twine taut and making sure it does not actually touch the bottoms of the cans, one of you talk softly into one can. Take turns speaking and listening.

You can test how well this works by getting another person without a can to stand the same distance away from both of you and ask if he or she can hear your conversation.

Sound travels about four times faster and farther in water than it does in air. This is why whales can keep in touch over huge distances in the sea, using a mixture of low, booming sounds and high-pitched squeals. Finback whales make long drawn-out sounds that help them communicate when almost 620 miles (1,000 km) apart.

18 Recording Sound

Sound was first recorded by a machine called a **phonograph** – a kind of early record player. A needle was attached to a drumhead, stretched across the narrow end of a sound horn. When someone shouted into the horn, the needle would vibrate. As the needle bounced up and down, it recorded the sound waves as grooves on a cylinder coated in wax or tinfoil.

Tape recordings
Sound recording techniques have come a long way since the days of the phonograph. Sound is now recorded onto magnetic tape. In the tape recorder, the sound waves are turned into electric impulses. These are stored on the tape as a sequence of different magnetic blips.

MAKE it WORK!
A microphone is a kind of electric ear. It turns sound waves into electric signals. However, it may have trouble picking up sounds not made close by. You can improve a simple microphone by putting it inside an umbrella! The umbrella's shape collects the sound waves and reflects them back to the microphone.

To play a phonograph recording, the cylinder was rotated underneath the needle. The pattern of bumps and dips in the cylinder grooves vibrated the needle, which the drumhead and sound horn then turned back into sound waves.

You will need
a tape recorder	blank cassette tapes
a microphone	earphones
tape	an umbrella
three or four friends who will sing	

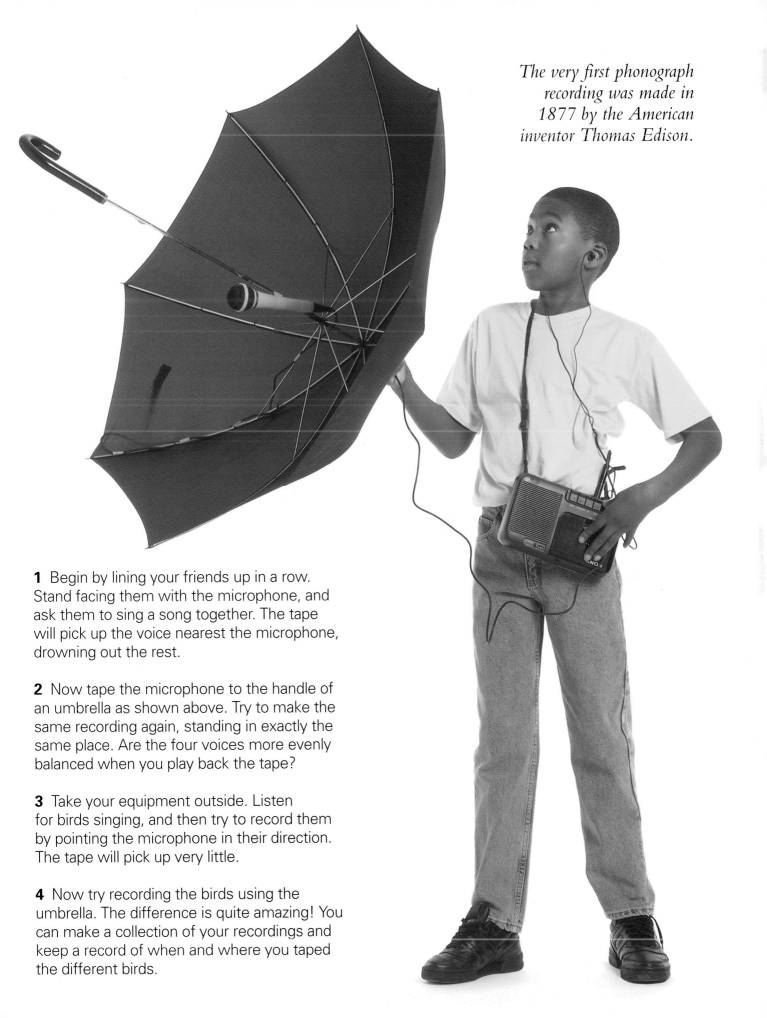

The very first phonograph recording was made in 1877 by the American inventor Thomas Edison.

1 Begin by lining your friends up in a row. Stand facing them with the microphone, and ask them to sing a song together. The tape will pick up the voice nearest the microphone, drowning out the rest.

2 Now tape the microphone to the handle of an umbrella as shown above. Try to make the same recording again, standing in exactly the same place. Are the four voices more evenly balanced when you play back the tape?

3 Take your equipment outside. Listen for birds singing, and then try to record them by pointing the microphone in their direction. The tape will pick up very little.

4 Now try recording the birds using the umbrella. The difference is quite amazing! You can make a collection of your recordings and keep a record of when and where you taped the different birds.

Bats have poor eyesight, but a very good sense of hearing. They can hear ultrasound – high-pitched sounds, way beyond the human hearing range. Bats can use ultrasound to find their way around the dark caves they live in. When a bat makes a high-pitched squeak, the sound bounces off the wall of the cave and returns to the bat as an ultrasound echo. From the amount of time between the squeak and echo, the bat can tell how far away the wall is.

1 Take a rectangle of black poster board, fold it over and draw the shape of a bat as shown above. Cut out this shape, unfold the card and you'll have a symmetrical bat. Make twelve bats, three for each player.

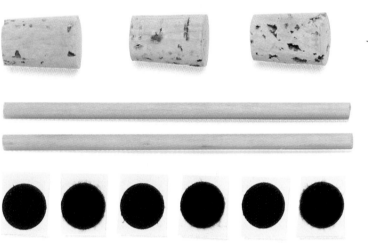

MAKE it WORK!
Make this batty ultrasound game! The playing board is a dark cave, criss-crossed by sound waves. The playing pieces are bats that cross the cave on the sound waves.

2 Stick a hook-and-loop fastener dot to the wide end of each cork, and then glue the side of the cork to the bat's body to make a kind of pedestal for each bat.

You will need
pennies or buttons for scoring counters
dots made of hook-and-loop fasteners, or paper clips
thick and thin colored poster board
several wooden dowels
a toothpick
12 corks
glue

3 If you don't have hook-and-loop fasteners or corks, simply attach a paper clip to the bottom of each bat as shown on the left. Hook-and-loop bats will sit on top of the sound waves. Paper clip bats, on the other hand, will hang below the sound waves.

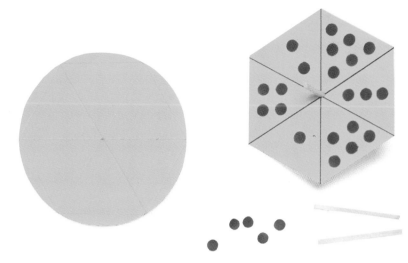

4 Cut dots from thin poster board, three each of four different colors. Put a different colored dot on each player's bats.

5 Make the spinner. Draw a circle and divide it into six segments. Snip off the sides to make a hexagon, and draw dots in each segement as shown.

6 Poke a toothpick through the middle of the spinner. Twirl it around and it will come to rest on one of the six sides.

Assembling the board

7 Now make the board. Cut a large square of thick poster board for the base. Cut four equal lengths of dowel and glue one upright in each corner.

8 Cut four strips of thick poster board, each as long as a side of the base board. Divide each strip into sixteen small squares. Glue these strips in place on the wooden dowels, as shown in the diagram above.

9 At each corner of the raised section put a square of paper in one of the players' colors.

10 Cut six long strips of thin poster board to represent sound waves. If you have made hook-and-loop fastener bats, mark off the sound waves with hook-and-loop fastener spots. If you have paper-clip bats, stick a row of colored-paper spots on each sound wave.

11 Put the sound waves in position, stretching the strips of thin poster board in wavy lines across the board from one raised side to the other. In some places, you will have to add extra dowel supports to hold up the sound waves and help them keep their shape.

Playing the ultrasound game

In this game, each player has three bats which set off from the corner marked in their color. The aim of the game is to make as many flights as possible from one side of the board to the other within a set time limit.

1 Decide on your time limit for the game – say fifteen or thirty minutes.

2 Spin the spinner. The player with the highest number starts, and play passes to the left.

3 The first player spins and moves a bat the corresponding number of spaces along the edge of the playing board. When a bat lands on a sound wave, it may cross the cave, spot by spot.

4 Each time a bat completes a crossing, the player puts a scoring counter in his or her corner of the cave. The bat continues around the edge of the board until it reaches another sound wave.

5 Players may use a spin to move any one of their bats. They may also split the score and move more than one bat.

6 When two players' bats going in opposite directions meet, they cannot pass one another. The two players both spin the spinner. The bat with the higher score continues its flight with a new spin, but the lower-scoring bat falls off and goes back to its corner.

7 At the end of the agreed time limit, the players have a race back to their corners, completing any waves they have already started. The first player to bring all three bats home gets three extra counters.

8 The winner is the player who has collected the most counters.

24 Radio Waves

Radio transmitters send sounds around the world, by changing sounds into electrical impulses and then **radio waves**. These waves of energy travel at the speed of light through air, solid objects, and empty space. Radio receivers pick up the waves and turn them back into sounds again.

MAKE it WORK!

Make your own simple radio receiver and try tuning into radio signals. Remember that the signals will be weaker than a normal radio.

Warning: This crystal radio does not need a power supply and must NEVER be connected to an electrical outlet.

You will need

two rubber bands
an awl
wood glue
paint
a sharp craft knife
wire strippers
22 ft. bare (non-insulated, non-lacquered) copper wire
a cardboard tube, 9 in. long
three nuts, bolts, and washers
33 ft. single-strand insulated electrical wire

thick balsa wood
16 in. steel wire
a large metal paper clip
a germanium diode, and a crystal earpiece from a hobby shop

1 Ask an adult to help you to cut a 9 in. x 4 in. piece of balsa wood for the base. Cut out four balsa wood feet and glue them to the corners.

2 Make two supports for the tube, by cutting out semicircles from the balsa wood. Glue into position on the base. Paint the wood.

3 Use the awl to make three small holes along the front of the base at **a**, **b**, and **c**. Put a washer over each hole. Push the bolts into the base and fix them under the board with nuts.

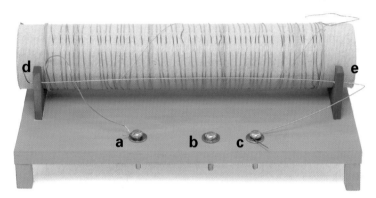

4 Take the bare copper wire and leaving one short end free at **e**, carefully wind the rest around the cardboard tube. The coils **must not** touch each other.

5 Secure the copper wire with rubber bands at both ends of the tube. Then wrap the long end of the copper wire around the bolt at **a**.

6 Push the steel wire through the balsa wood supports in front of the copper coil, from **d** to **e**. Leave a short end at **d**, pushing it upward to secure. Bend the long end at **e** forward and wrap it around the bolt at **c**.

▲ The **diode** detects radio waves picked up by the **antenna** so that you can hear sounds in the earpiece.

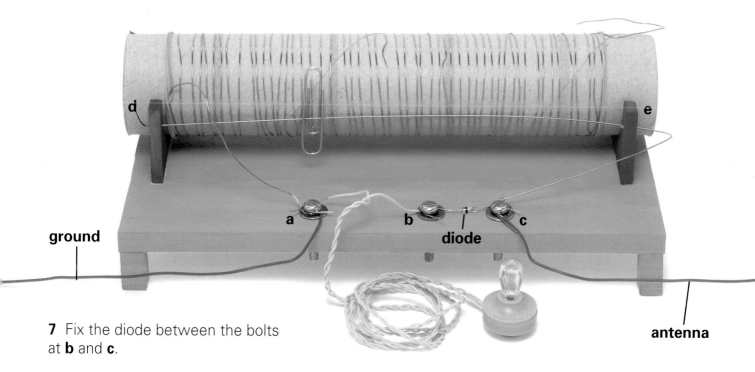

ground

a **b** **c**

diode

antenna

7 Fix the diode between the bolts at **b** and **c**.

8 Strip the insulation from the ends of both earpiece leads, using the wire strippers. Attach one end to the bolt at **a**, so that it touches the wire. Wrap the other end around the bolt at **b**.

9 To make an antenna, cut 30 ft. of insulated wire. Strip off the insulation at one end of the wire and tie it around the bolt at **c**. Tie the other end to a post or tree in an outside space.

10 To ground the radio, take 3 ft. of insulated wire and strip the covering from both ends. Attach one end to the bolt at **a**, and the other to a metal object, such as a clean, unpainted metal railing.

NOTE
1. Keep surfaces of all connections clean.
2. Make sure all connections touch in the correct places, and nowhere else.
3. If you use lacquered copper wire, you must sandpaper the ends where they wrap around the bolts, and where the paper clip touches the copper wire.
4. Weather and location may affect reception. Your receiver will probably work best in a large open space away from tall buildings.
5. Try using different objects as antennas, and to ground your radio.

▼ To operate the radio, put in the earpiece and slip a metal paper clip onto the steel wire between **d** and **e**. Twist it backward, and move it slowly along the copper coil. You should hear faint clicks or a radio station. What the radio picks up depends on the number of turns of copper wire.

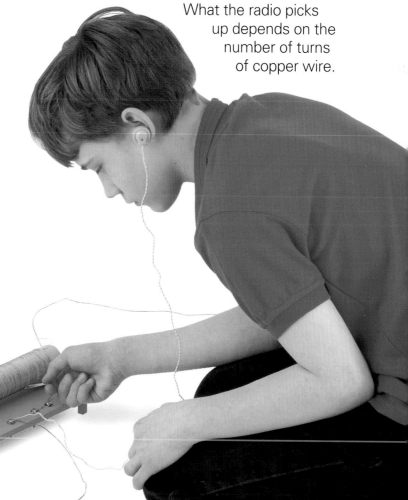

Every sound has its own pitch – high or low – depending on the shape of the sound wave. To a musician, each different pitch of sound makes an individual musical **note**.

Musical notes can be arranged in a special pattern called a **scale**. There are many kinds of scales, but the most common is called an **octave**. It is made up of eight notes that move higher and lower step by step. The notes on a piano are set out in octaves – and you might have sung the notes of an octave yourself: do, re, mi, fa, so, la, ti, do.

MAKE it WORK!

Make your own instruments that play the notes of the octave scale. You will need to learn the notes of the octave, or copy them from a musical instrument – an instrument with fixed notes, such as a piano or electric organ, is easiest. If you don't play an instrument yourself, you could ask for some help from a friend or adult who does.

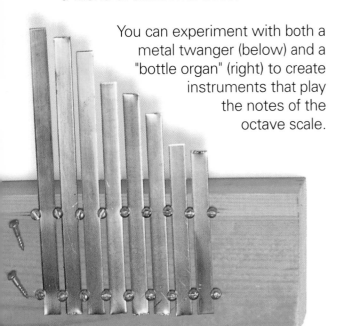

You can experiment with both a metal twanger (below) and a "bottle organ" (right) to create instruments that play the notes of the octave scale.

For the "bottle organ" you will need

food coloring   a long nail
a musical instrument   water
eight bottles of the same size and shape

1 Collect eight identical bottles and set them up in a line. (Milk bottles are probably the easiest to use.) Hit each bottle with a long nail. They should all make the same sound.

2 Take a jug of water and pour a little into the first bottle. If you are going to copy a scale from a musical instrument, play the first note of the scale. Then gently hit the top of the bottle with the nail to see if it sounds close to the note you have just played. Gradually adjust the water level in the bottle until both bottle and instrument give out exactly the same note.

3 Follow the same method for each bottle until you have made a complete octave.

4 When you are happy with the sound of the "bottle organ," dye the water in each bottle with a few drops of food coloring. You will be able to see the different notes more easily.

5 Now try to pick out a tune – something simple, such as "Twinkle, Twinkle, Little Star."

Making a "metal twanger"

You can make another octave scale using strips of brass (previous page) or thick steel wire (right) fixed to a solid block of wood.

For a "brass twanger" you will need

a small block of wood   a file
a thin strip of brass   screws
a small hacksaw   a screwdriver
a bradawl   a hammer

Be careful! To make a "twanger," you will have to use sharp tools. Ask an adult to help you.

1 Cut a short strip of brass and hold it in place on the block of wood. Pluck the end of the strip, and it will make a twanging noise.

2 Now cut seven other strips of brass, each slightly longer than the last. Experiment with different lengths until you can arrange the strips on the block of wood to make an octave.

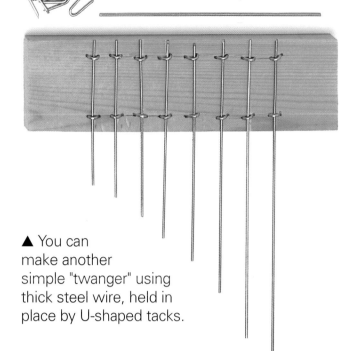

▲ You can make another simple "twanger" using thick steel wire, held in place by U-shaped tacks.

3 File the sharp edges off the strips of brass and fix them in place. Mark where the screws should go with a bradawl, and then twist them into the wood with the screwdriver.

▼ Musicians check the pitch of an instrument with a tuning fork. When the fork is struck, its prongs vibrate at a regular speed, sending out a clear, single note that never changes.

A lot of the pop music and classical music we hear is based on the eight-note scale, but some cultures use totally different scales. Much Chinese music is based on a sequence of five notes; Indian classical music often uses a twenty-two note scale.

How many kinds of musical instruments can you think of? They come in many different shapes and sizes, from guitar to piano, synthesizer to tom-tom drums. Yet they all have one thing in common – they create sounds by making air vibrate.

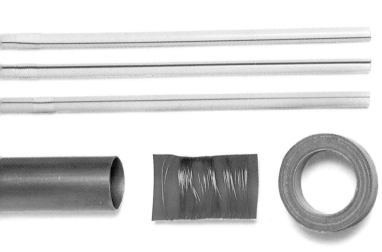

Musical pipes

Many musical instruments have pipes. Inside each pipe is a column of air. When this air is made to vibrate in a certain way, the pipe plays a note.

MAKE it WORK!

Make your own music using pipe instruments. Panpipes are a set of pipes of different lengths, each of which makes a different note. Tubular bells are metal pipes that hang free and make a bell-like sound when hit with a large nail.

To make the panpipes

1 Have an adult help you use the utility knife to cut eight straws or eight pieces of plastic tubing in different lengths. Tape them together, stepped at one end, as shown at the right.

2 To play the panpipes, blow gently across the top of each pipe at the end that is secured with tape to produce a note. You will find that longer pipes play a lower note than shorter pipes.

You will need

glue
wood
copper pipe
nylon fishing line
flexible plastic tubing
screws and a screwdriver
plastic drinking straws

tape
utility knife
a hacksaw
a large nail

3 Experiment with the lengths of the pipes. Can you make octave scales like those on the previous page?

To make the tubular bells

1 Ask an adult to help you cut eight pieces of copper pipe. Each one should be a little longer than the one before, so that you can arrange them in a stepped pattern.

4 Hang the copper pipes from the top piece of the frame using pieces of strong fishing line. Thread the line through the holes you have drilled, and attach it to the copper pipes with tape.

Pipes that give out a sound when they are struck are **percussion instruments**. *Pipes that are blown to make music are* **wind instruments**.

2 Make a wooden frame. It should be big enough to hold the copper pipes hanging at well-spaced intervals. You will need a flat base, two side pieces and a top piece.

3 Drill eight holes at equal spaces through the top piece of the frame. Then put the frame together, cutting the side supports into the baseboard. Glue the joints and secure them firmly with screws.

5 Arrange the pipes in order of length. Strike them with a metal nail to make them chime.

Because the pipes hang free, they continue to vibrate after they have been hit, and thus, the sound lingers on. Try striking a pipe and then grasping the pipe tightly with your hand. The sound dies right away. As soon as you stop the pipe from vibrating, the sound waves stop being produced too.

Unlike our panpipes, many instruments create sounds by vibrating the air in just a single pipe. In these instruments, the different notes are made by altering the length of the main pipe. Recorders have holes to let the air out in different places. A brass trombone has a slide that moves up and down to change its length.

1 Take a piece of copper pipe or a section of cardboard tubing. Leave a space at one end, and then use a bradawl to mark out evenly spaced holes along the rest of the pipe.

2 Ask an adult to drill the holes you have marked. At the end of the pipe nearest the mouthpiece, cut a blowhole, following the diagram on the right.

▼ metal recorder

▼ brass-pipe recorder

▼ cardboard-tube recorder

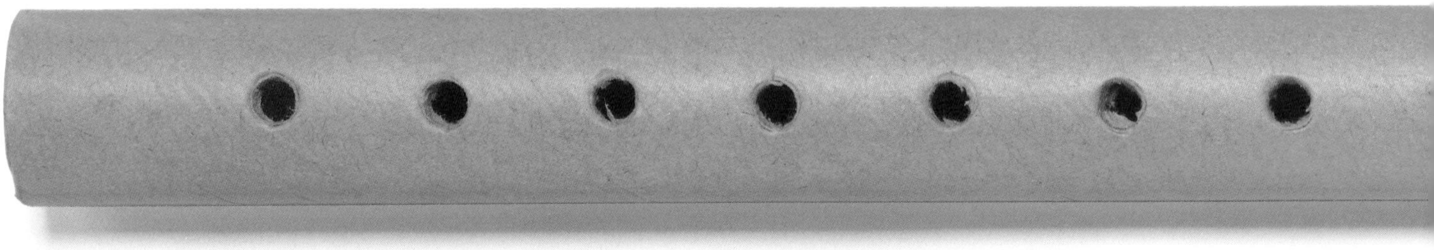

▲ ▼ slide whistles

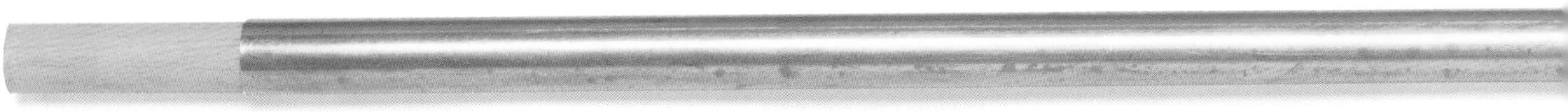

Make a recorder
A recorder player sounds different notes by putting fingers over the holes to make the air travel different distances down the pipe.

To make a recorder you will need
a tube of thick cardboard or copper pipe
a bradawl, drill and hacksaw
a cork

3 Cut a slice off a piece of cork to give it a flat edge. Then sandpaper the cork to make it smooth and fit snugly inside the top of the recorder pipe. Position the flat side face up, opposite the blowhole and finger holes.

4 Now blow gently into the mouthpiece. The cork and the blowhole are shaped so that the air inside the recorder is vibrated.

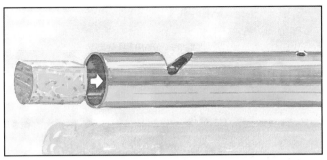

Making the recorder mouthpiece

To make a slide whistle you will need

a copper pipe   tape or glue
wooden dowel
the mouthpiece from a party blower

1 Ask an adult to cut a piece of pipe.

2 Cut a piece of dowel about 6 inches longer than the pipe. The dowel should fit inside the pipe as tightly as possible, while still being able to move easily.

3 Fit the party blower mouthpiece to one end, using glue or tape. If you cannot find a mouthpiece that fits, you can still play the slide whistle by blowing across the top of the pipe.

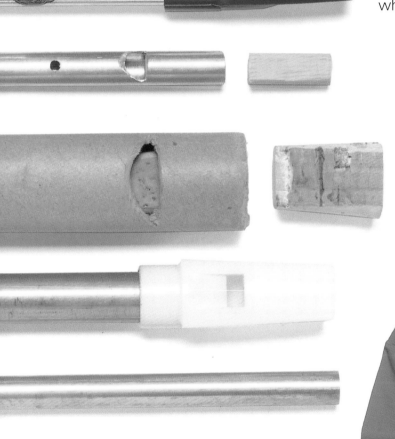

Make a side whistle

Slide whistles make a weird wailing sound. The length of the pipe is changed by pushing a dowel in and out. There are no separate notes, just a continuous rising and falling tone.

The width of a pipe, as well as its length, will affect the sound it makes. Narrow pipes make higher-pitched notes than wide pipes.

Shakers and rattles are percussion instruments. They are often used to stress the rhythm of a piece of music and give it a strong beat. Shakers, like our paper cup shakers, make the air vibrate by moving loose objects inside an enclosed space. Rasps and rattles rub two rough surfaces together to make sound vibrations.

To make the shakers you will need

paper cups   tape
plastic bottles with lids
rice, lentils, chickpeas, beads and pebbles

1 To make a paper cup shaker, put a handful of rice or lentils into one cup. Turn another cup upside down and tape the two cups together, rim to rim.

2 To make plastic bottle shakers, simply pour a handful of beads or chickpeas into the bottle, and put the lid on tight. You could decorate the shakers with colored paper if you wish.

3 Try making shakers with different-sized bottles. You will find that larger bottles which hold more air make deeper sounds.

MAKE it WORK!

Make your own collection of different shakers and rattles. They can make a variety of sounds, depending on the amount of air each one vibrates and the kinds of surfaces that are rubbed up against one another.

▲ Experiment with different fillings for your shakers. You will find that paper cup shakers with lentils make a softer sound than plastic bottles with chickpeas.

To make a rattle you will need

a wooden dowel or stick   a hammer
several long, thin nails   enamel paints
metal bottle tops   a larger nail

1 Punch a small hole in the middle of each bottle top using the hammer and larger nail. **Be careful** as you do this! Put a piece of old wood underneath the bottle top, and don't hit your fingers.

2 Slip four bottle tops onto each thin nail. (If you like, paint them first.) Then hammer the nails into the dowel or stick, making sure the tops can rattle freely, and that they don't slip off the ends of the nails.

To make a sandpaper rasp you will need

two blocks of wood   thumbtacks
two sheets of sandpaper

1 Tack the sandpaper to the blocks of wood as shown below.

2 To play the rasp, rub the two sandpaper surfaces together. It makes a soft, grating noise.

To make a wooden rasp you will need

a block of soft wood   a large nail
a small hacksaw

1 Ask an adult to help you cut a zig-zag shape along the top of the block of wood as shown.

2 To play the wooden rasp, run the nail back and forth along the uneven surface. It makes a harsher sound than the sandpaper rasp.

Both of these rasps make soft noises. Unlike the shakers, they have no space inside them where the air can vibrate to make the sounds louder.

34 Drums

Drums are probably the oldest and simplest musical instruments in the world. They contain a space filled with air, and have a flexible drumhead stretched across one end. When the drumhead is struck, it vibrates and makes a noise.

Although all drums work in basically the same way, they are able to make a range of different noises. Size is important. A big bass drum makes a much deeper sound than a small bongo drum. The drum's pitch is also affected by the drumhead. A tight drumhead makes a higher note than a slack one.

MAKE it WORK!
Experiment with the sounds and tones of different drums. Try out some of these ideas and, if you like, put together a collection of drums to make your own drum kit.

You will need

balloons	thick paper
string	rubber bands
glue or tape	thin wooden dowels

boxes and tin cans of all shapes and sizes
a sheet of plastic or an old plastic bag
an eyelet punch or cardboard hole reinforcers

▶ Bongo drums
1 Cut cardboard tubes into several different lengths, to make bongo drums that will sound different notes.

2 Cut flat pieces of balloon rubber to make the drumheads, and secure them across the tops of the cardboard tubes with elastic bands.

3 Attach a dowel to each drum. That way you can hang them up, and the notes will sound more clearly from the open ends of the drums.

▲ Tin can drums
1 Take the top and bottom off a tin can. Wash the can, being very careful not to cut yourself on any sharp edges inside.

2 Stretch balloon rubber over the ends of the can, and secure it with elastic bands.

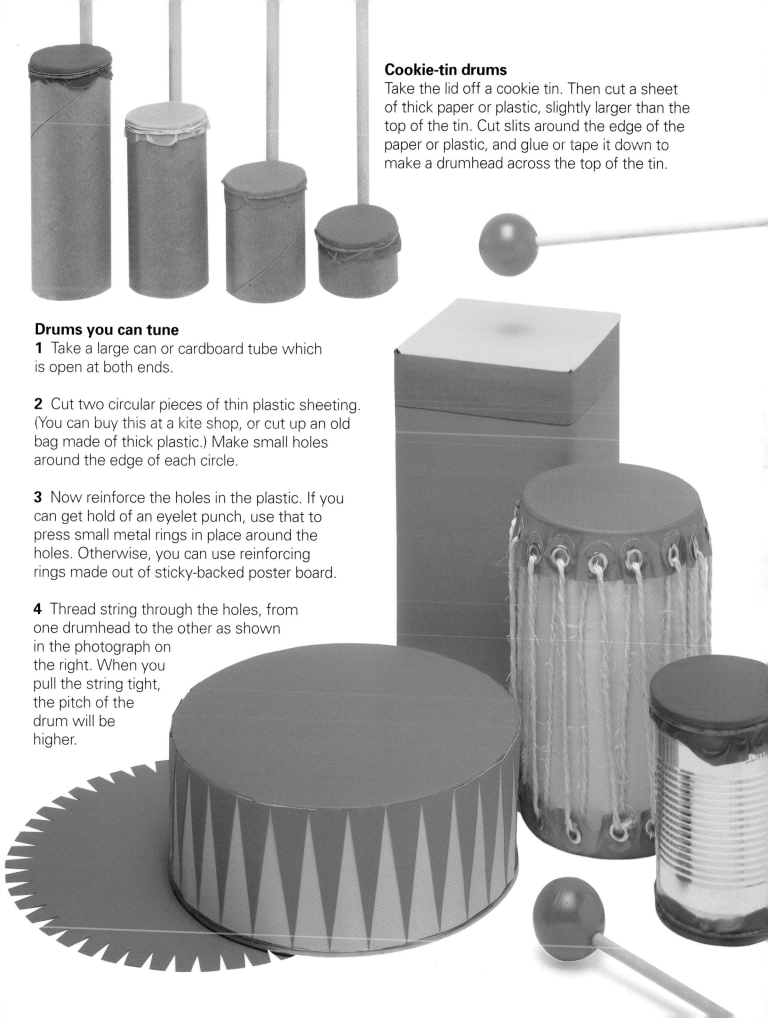

Cookie-tin drums

Take the lid off a cookie tin. Then cut a sheet of thick paper or plastic, slightly larger than the top of the tin. Cut slits around the edge of the paper or plastic, and glue or tape it down to make a drumhead across the top of the tin.

Drums you can tune

1 Take a large can or cardboard tube which is open at both ends.

2 Cut two circular pieces of thin plastic sheeting. (You can buy this at a kite shop, or cut up an old bag made of thick plastic.) Make small holes around the edge of each circle.

3 Now reinforce the holes in the plastic. If you can get hold of an eyelet punch, use that to press small metal rings in place around the holes. Otherwise, you can use reinforcing rings made out of sticky-backed poster board.

4 Thread string through the holes, from one drumhead to the other as shown in the photograph on the right. When you pull the string tight, the pitch of the drum will be higher.

In England whirling rattles like the one on this page were a common sight at soccer games. Fans on the sidelines would wave their rattles to cheer their team on.

Our rattle makes a very loud noise. It creates sound in two different ways. The wooden surfaces of the cog and tongue make a clattering noise as they hit one another, while the whirling movement of the rattle sets off its own rhythmic pattern of sound vibrations in the air around it.

You will need

four metal washers	a thread spool
a drill and hacksaw	strong wood glue
thin wooden dowel	
thick wooden dowel	

a fairly thick piece of wood about 1/4 in. thick for the frame of the rattle
thinner wood for the tongue of the rattle
matchsticks or thin slivers of wood

Be careful! This project involves difficult woodworking. Ask an adult to help you with all the stages when you have to cut the wood or drill holes.

MAKE it WORK!

Make your own whirling rattle, take it out of doors and see how loud a noise you can make with it!

top and bottom of rattle frame

rattle tongue

thin dowel

pieces for end of frame

thread spool for cog

slivers of wood   washers   thick dowel

1 Cut two identical rectangular pieces of the thicker wood. These will form the top and bottom of the frame of the rattle.

2 Cut a rectangle of thin wood, slightly shorter than the width of the frame. This will be the tongue of the rattle.

3 Cut the thick dowel in two – you need a long piece for the handle of the rattle, and a much shorter piece for the top.

4 Cut two small rectangles of thick wood. They fit across the outer end of the rattle, with the tongue sandwiched between them.

5 Make the cog wheel. Cut several thin slivers of wood, or cut the heads off several matchsticks. Then glue these sticks around the side of the spool, using strong wood glue.

6 Take a drill the same width as the thin wooden dowel. Drill a hole through one end of the top and bottom pieces of the frame. Then drill the same size holes into the center of both pieces of thick wooden dowel.

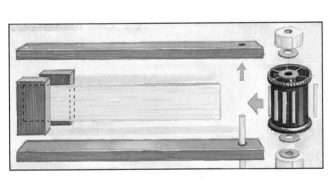

Assembling the whirling rattle

Be careful! Make sure that you use very strong wood glue to stick the pieces of the rattle together. Follow the instructions on the package carefully, and try not to let the glue touch your skin.

7 Assemble the rattle as shown in the drawing below. Thread the thin dowel through the cog and slip a washer on either side. Then add the top and bottom of the frame, followed by two more washers. Fit the ends of the thin dowel into the holes you have drilled in the thicker dowel.

8 Now glue the tongue and the outer end of the rattle in place. You must position the tongue very carefully. It should just touch the cog, so that it makes a noise, but should still be springy enough to let the cog spin around freely.

The washers in the whirling rattle reduce the friction between the pieces of wood and help the rattle swing more easily.

Have you ever burst a balloon by accident? The loud, unexpected bang can make you almost jump out of your skin! The air trapped inside the balloon suddenly rushes out, creating sound waves which we hear as a loud bang. The sound of any explosion is caused by air moving at great speed – that's why the blast from a bomb will knock over people and even buildings.

MAKE it WORK!

Moving air will make all sorts of noises. It can hum, whistle and whir, as well as bang. Put together these simple cardboard gadgets and see how they create sounds that crash like a thunder clap or hum like a bumble bee.

You will need

colored poster board   glue
brown wrapping paper   string
scissors or utility knife   a metal washer

Making a "banger"

1 Cut a square of poster board.

2 Cut another square of brown paper, slightly larger. Snip this square in half from corner to corner to make a triangle.

3 Place the brown paper on top of the poster board as shown, and fold it where it overlaps. Glue paper and poster board along the overlaps.

4 Fold both poster board and paper from corner to corner.

5 To make a sound, hold the corner of the "banger" as shown on the left. Sharply fling it downward. The paper beak will flick out and make a bang.

▲ Making a "whirrer"

1 Cut out the two cardboard shapes shown on the left, to make the wing and flapper sections. Slot them together.

2 Pierce two holes in the wing as shown, and glue a small metal washer to the nose to weight it down. Thread about a yard of string through the two holes.

3 Take the "whirrer" outside and spin it around your head. At first, the wing will just make a clattering noise, but as you speed up the sound will change to a high-pitched whir.

▼ Making a "spinner"

1 Cut two hexagons of colored poster board. Make two small holes at the center, and four or five larger holes around the edges.

2 Glue the hexagons together. Thread a piece of string through the center holes and tie the ends together.

3 Twist the string around and around. Then pull it outward. As you pull, the string will keep on winding and unwinding itself and the spinner will keep on humming.

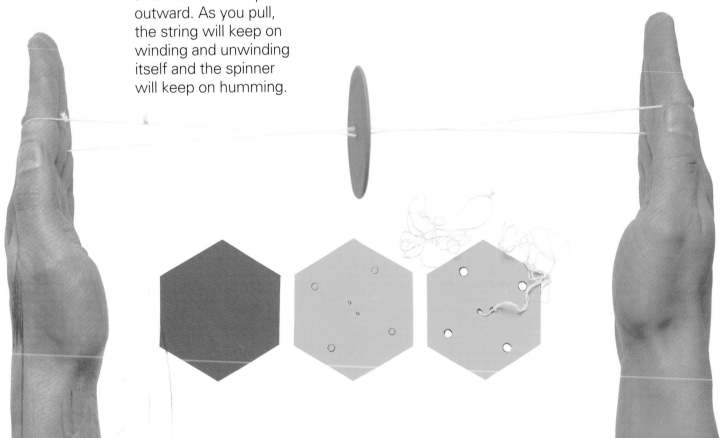

String instruments make music when a string or cord is plucked, so that the air around the string starts to vibrate. The pattern of a string instrument's sound waves depends on three things: the length of the string, what the string is made of and how tight it is.

A vibrating string on its own does not make much noise, so to make their sound louder, most string instruments have a **soundboard** and **resonator**. The soundboard picks up the strings' vibrations and transfers them to the resonator – a big space filled with air, which amplifies the sound.

You will need

a wood box
a broom handle
a hammer
string
a drill
nails

Make a wood box bass

Make your own instrument with one adjustable string and a wood box resonator.

1 Ask an adult to help you drill a hole in one corner of the wood box. The broom handle should fit loosely inside the hole.

Drilling the hole   Fitting the handle

2 Tap nails into the opposite corner of the wood box and the top of the broom handle. Tie a piece of string between the two nails.

3 Put one foot on the box to hold it steady and pluck the string. Pull the handle back to tighten the string and make higher notes. Push it forward for lower notes.

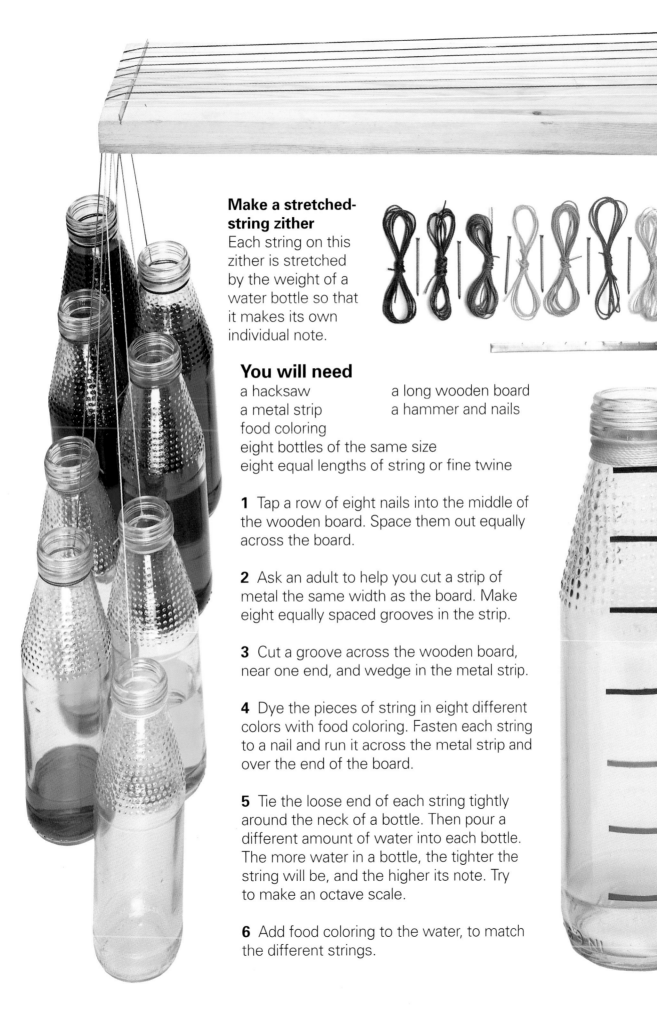

Make a stretched-string zither

Each string on this zither is stretched by the weight of a water bottle so that it makes its own individual note.

You will need

a hacksaw
a metal strip
food coloring
a long wooden board
a hammer and nails
eight bottles of the same size
eight equal lengths of string or fine twine

1 Tap a row of eight nails into the middle of the wooden board. Space them out equally across the board.

2 Ask an adult to help you cut a strip of metal the same width as the board. Make eight equally spaced grooves in the strip.

3 Cut a groove across the wooden board, near one end, and wedge in the metal strip.

4 Dye the pieces of string in eight different colors with food coloring. Fasten each string to a nail and run it across the metal strip and over the end of the board.

5 Tie the loose end of each string tightly around the neck of a bottle. Then pour a different amount of water into each bottle. The more water in a bottle, the tighter the string will be, and the higher its note. Try to make an octave scale.

6 Add food coloring to the water, to match the different strings.

Playing strings

Many string instruments call for nimble fingers. Violinists or guitar players, for instance, hold down the strings in different places along the necks of their instruments. By altering the length of the strings they make different notes.

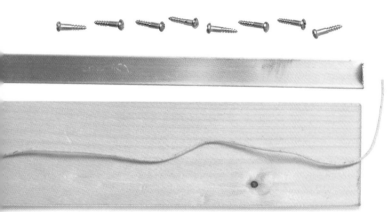

Make a banjo

This homemade banjo is almost like the real thing. It has a bridge to transfer the strings' vibrations to the soundboard, and a large, plastic tub which acts as a resonator. See how many different notes it will play.

You will need

a large, plastic tub
a strip of metal
wood, glue, and string
a bradawl and hacksaw
a screwdriver and eight screws

1 Cut a piece of wood for the soundboard. Its length should be about three times the width of the plastic tub.

2 Ask an adult to cut four small grooves in the metal strip, to make a **bridge**. Then cut another groove across the soundboard, near the bottom, and glue the bridge in it.

3 Mark holes in the wood where the screws will fit, following the photograph below. Twist the screws down a little way, and tie on the strings, leading them across the metal bridge.

4 Ask an adult to cut another piece of wood, the same depth as the plastic tub, to support the soundboard. Glue this support and the soundboard in place.

5 Twist the screws down farther in order to tighten up the strings and fine tune the banjo.

Matchbox guitar

Stretch four rubber bands around a matchbox as shown on the left, and insert a bridge of thick poster board. The angle on the bridge means that each band is stretched by a different amount, so that it sounds a different note.

Making guitars

Rubber band instruments are easy to make. Try either a large shoebox guitar, or the smaller, portable matchbox version.

Shoebox guitar

To make the shoebox guitar, just cut a hole in the lid, and stretch the rubber bands across. Make the bridge from a sandwich of thick poster board and colored paper as shown below. It should be strong enough that you can slide it along to make the strings tighter.

You will need

a shoebox
matchboxes
a utility knife or scissors
rubber bands of all sizes
thick poster board and colored paper

The shoebox guitar makes a deeper, richer sound than the little matchbox models, because the resonator is bigger and the strings (the rubber bands) are longer.

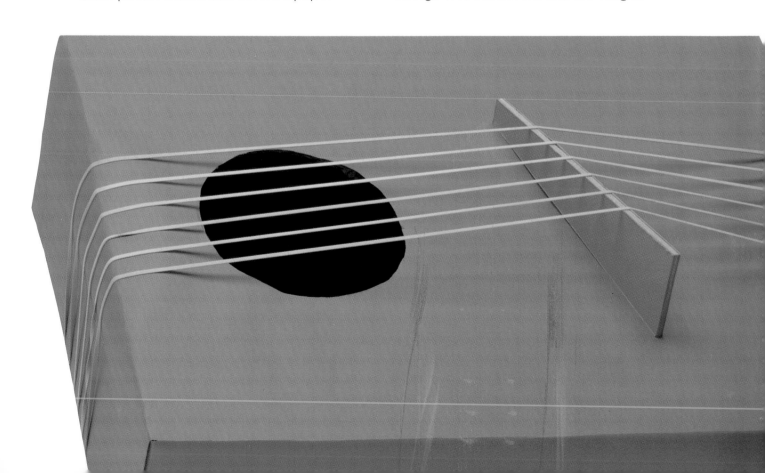

Nowadays, we can record music on cassettes or compact discs, but for hundreds of years, the only way of recording music was to write it down on paper. Over the centuries, a standardized system for writing music developed. This system of **musical notation** is now used for western music in many different countries.

The symbols and instructions on a **score** give all kinds of information to the musician – which notes to play and what rhythm to follow, how loud or soft, and how fast or slow the piece is. Learning to read and write music is like learning a new language.

▼ Western music is usually written on a set of five lines and four spaces called a **staff**. The positions of the notes on the staff tell the musicians which notes to play.

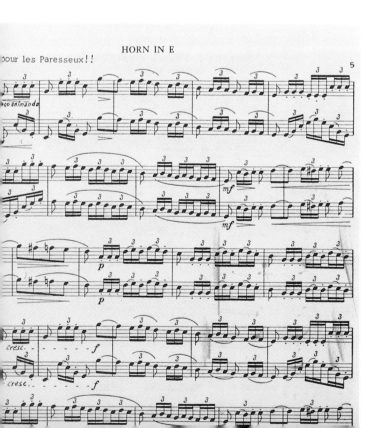

Some of the instructions on a musical score – especially those about the speed of a piece – are usually written in Italian. Andante, for instance, means slowly. Prestissimo means as fast as you can.

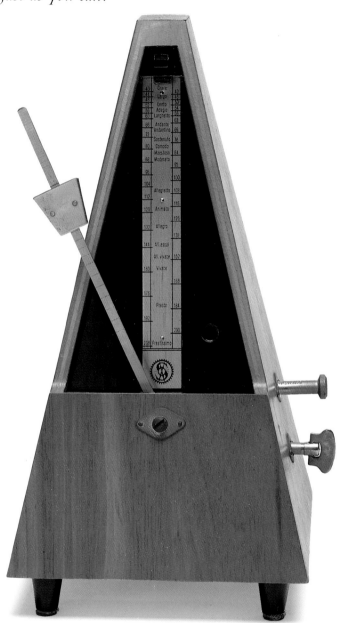

▲ To check that they are playing a piece at the right speed, musicians use a clockwork timer called a metronome. It has an upright pendulum that ticks backward and forward with a regular beat. By moving the weight up or down the pendulum, it can be made to tick more slowly or quickly.

MAKE it WORK!

You can work out your own simple system of writing down music and use it to record any music you compose for your homemade instruments.

1 Color in the stick-on spots to match the colors of the stretched-string zither on page 41 or the bottle organ on page 26. The different colored spots will stand for the different notes.

2 The size of the spots tells you how loudly to play the instrument. Small spots mean you play the note softly *(piano)*, medium-sized spots mean medium loud *(mezzo forte)* and large spots are loud *(forte)*.

You will need

graph paper
white spot stickers in three sizes

▶ Here's a simple example, going up a scale and down again, not too loud and not too soft.

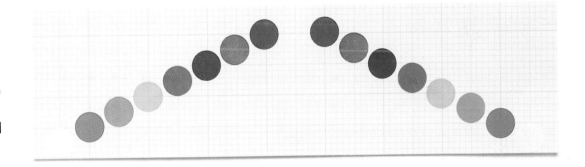

▶ Up and down a scale again. Softly on the way up, loudly on the way down.

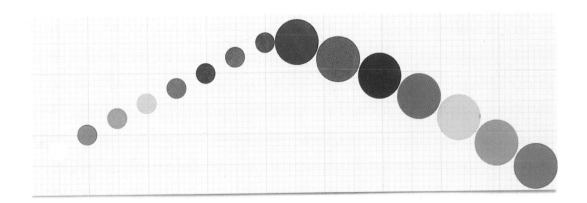

▶ Music for two players. The line stands for a drum beat every other note. The guitar plays orange and red notes, getting louder and louder.

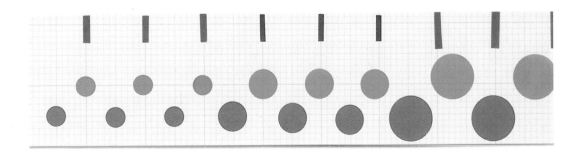

Amplify When you amplify something, you make it louder.

Antenna A long piece of wire or metal that picks up radio waves. All radios and televisions need an antenna.

Bridge In a string instrument, the bridge is a small, raised piece of wood or metal that connects the strings to the soundboard.

Cochlea A curly tube inside the ear, filled with liquid and lined with tiny hairs. It turns the vibrations made by sound waves against the eardrum into nerve pulses.

Decibel Decibels are the unit for measuring how loud or soft a sound is.

Diode A tube-shaped device used in electronic circuits. The diode prevents the current from flowing both ways through the circuit and makes it flow in just one direction.

Ear drum The part of the ear inside the head that vibrates when struck by sound waves.

Hertz Sound waves are measured in a unit called the Hertz. It is named after the German physicist, Heinrich Hertz.

Incus One of the three main bones inside the ear. Often also called the anvil bone.

Malleus One of the bones inside the ear. It is often also called the hammer bone.

Musical notation Recording music by writing it down on paper.

Nerve pulses Nerve pulses are signals sent out by our organs of sense, such as our eyes and ears. These signals pass along our nerves, to the brain, where they are decoded as sights, sounds, tastes, smells or feelings.

Note In music, a note is a sound played at a particular pitch.

Octave An octave is a scale of eight notes that rise and fall step by step. Most Western music is based on this octave scale.

Percussion instruments Musical instruments that are played by being struck or shaken. Drums, maracas and xylophones are all examples of percussion instruments.

Phonograph An early kind of record player. It recorded sounds by making a series of bumps and dips in grooves on a rotating cylinder.

Physics The branch of science that finds out about different kinds of energy and matter. Sound is one of the forms of energy investigated by physicists.

Pitch The pitch of a note is how high or how low it is. Different-shaped sound waves make notes at different pitches.

Radio waves Waves of electromagnetic energy that travel rapidly across long distances. Sounds are turned into radio waves by a radio transmitter, and changed back into sound again by a radio receiver.

Resonator A space inside an instrument filled with air, which vibrates and makes the sound of the instrument louder.

Scale A series or sequence of different musical notes.

Score A written record of a piece of music.

Sound waves Sounds travel through the air in waves. When someone bangs on a drum, for example, the drumhead squashes together the air beside it every time it vibrates. If we could see air, the movement of sound waves would look something like the ripples on a pond when a stone is thrown in the water.

Soundboard The board in a piano or a string instrument that is connected to the strings by the bridge. It vibrates when the strings are struck or plucked.

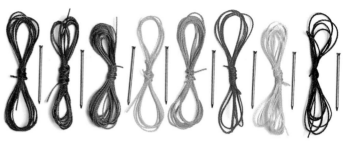

Staff In music, a staff is a set of five lines on which the notes are written.

Stapes One of the bones of the inner ear. It is often also called the stirrup bone.

Stethoscope An instrument used by a doctor to listen to the inside of a patient's body.

String instruments String instruments are played by plucking strings or scraping them with a bow. A guitar, sitar, and zither are all examples of string instruments.

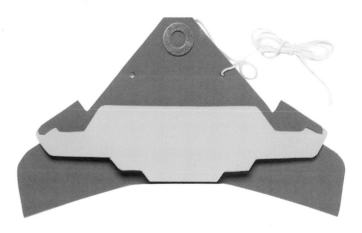

Vacuum A completely empty space that does not even contain air. Outer space is a vacuum.

Vibrate When something vibrates, it moves back and forth a small amount very quickly. A guitar string vibrates if you pluck it.

Wind instruments Instruments that are played by blowing through them. Trumpets and recorders are both wind instruments.